More on the Well

Copyright © by John Swofford 2024

1

We'll take the energy of a finite square well:

$E_n = \frac{\hbar^2 \pi^2 n^2}{2ma^2}$, where $k = \frac{n\pi}{a}$; we'll plug in the

mass of an electron, $(9.109383702 * 10^{-31})$, and

solve for E_1 . Where a is the breadth of the well,

and $a = 2\pi$, and $n = 1$, we have $E_1 =$

$1.526066079 * 10^{-39}$.

$\frac{E_1}{m_1} = 1.675268194 * 10^{-9}$.

The kinetic energy (E_1) is one half the mass times the velocity squared, and the gravitational potential is v^2 . Energy in excess of the gravitational potential would generate an electric potential and or an amplitude above the well: (mass associated with the potential would be positively or negatively charged depending on the charge of an associated scattered state—a wave of incoming energy crossing the region above the well). So, things can become charged (or change their helicity) when they take on or interact with negative kinetic energy or bound states. Mass in a well, then, would be a function of charge outside the well and vice versa.

Now, as such, velocity in the well that would be in excess of the speed of light becomes gravitational potential inside the well—and it adds to an electric potential above the well. Amplitude outside the well, then, would be the maximum field strength of the electric field. Photons, in this way, conduct electric and magnetic energy.

I take $\frac{E_1}{m_1}$ to be one half the square of the velocity in the well and or the range of that which lives inside the well and may produce particles of lesser or greater mass than an electron. That means I'm taking the mass that would exist outside the well and defining it as a function of velocity squared or gravitational potential.

The mass, then, is potential in excess of the velocity squared. To produce a particle, the velocity must exceed the value of this total potential; we divide J (joules) by v^2 and that leaves us with kilograms; in our case, that's the mass of an electron.

We have $\frac{k}{v} = \frac{m}{\hbar}$ and so we have $\frac{\frac{n\pi}{a}}{v} = \frac{m}{\hbar}$ and thus we have $v_1 = 5.788381802 * 10^{-5}$. That's the velocity of the electron inside the well.

If the energy (on both sides of the well—where $E = mc^2$) exceeds the value of v^2 , then we can expect the electric field to propagate.

We'll hold that energy (in the well) to be constant; therefore, as the energy or velocity in the well would increase, things can happen outside the well; we can plug various masses into our equation and solve for what n must be; (bound states would require bound (integer) steps for n ; I'll argue, however, that fractions of n correspond to electric potential and or electric field strength outside the well).

Let's return to the energy of the bound state:

$E_n = \frac{\hbar^2 \pi^2 n^2}{2ma^2}$. If $k = 10430.37835$, (an arbitrary number), then, where $E = \hbar c k$, we have $E_n = \frac{1}{2} * 3.297592039 * 10^{-22}$ (where V would equal the other half).

We have:

$$n^2 = \frac{(9.109383702*10^{-31})(3.297592039*10^{-22})(2\pi)^2}{(1.054571817^{-34})^2(\pi^2)}.$$

So, $n = 3.286977987 * 10^8$.

That's the value of n when n is applied to the mass an electron.

We know that $k = \frac{n\pi}{a}$, and we define k (in a normalized state) to be increasing by factors of two as the amplitude decreases: (in a non-normalized state—where the amplitude is held constant—k would be decreasing by factors of two as the wavelength increases): so we can write the change in k to be: $\log_2 n = 28.292186554$.

Now, n must be an integer (corresponding to intervals of π —where the sine function would be equal to zero, and, over the breadth of the well, a , forming a half wave and then twice that and then twice that and so on) so we have 27 intervals of n and $28^{.292186554}$ extra intervals of k which correspond to joules and or an electromagnetic potential of joules per coulomb.

Now, we take $28^{.292186554} = 2.647525296$ (for an electron).

We assume that k corresponds to an electric field—it is an electromagnetic wave in this case—and so the wave function, as an electromagnetic wave, must have an electric potential of

$1.341057427 * 10^{-44}$ J/C . (That was $\hbar c k / C$).

Next, we'll plug in the mass of a muon, (which has the same charge and spin of an electron): the mass of a muon is: $1.883531627 * 10^{-28}$; n must be $4.726490141 * 10^{9}$; $\log_2 n = 32.138122100$; we arrive at an electric potential of $8.175263007 * 10^{-45}$.

Next, we'll plug in the mass of a tau, (which has the same charge and spin of an electron): the mass of a tau is: $3.16754 * 10^{-27}$; n must be $1.938265884 * 10^{10}$; $\log_2 n = 34.174047437$; we arrive at an electric potential of $9.357444023 * 10^{-45}$.

We have: $E_n + V = mc^2$; when it comes to an electron, that yields a positive potential (outside the well) of $8.187105761 * 10^{-14}$ and an electric potential (of an electric field outside the well) of $1.311718955 * 10^{-32}$.

The electron wave function generates its own electric potential, as calculated in the above; we divide the former by the latter, and we arrive at a constant: $1.022366431 * 10^{-12}$. That tells us the overall electric potential of the wave packet's electric field existing outside of the well.

The potential outside the well, V , when there is a muon in the area, is: $1.692833804 * 10^{-11}$. We divide by a coulomb and we have an electric potential and or field (in the area) of $2.712218766 * 10^{-30}$. The muon wave function generates its own electric potential: we divide the electric potential of the muon wave function by the electric potential of the electric field outside the well and we have $3.014234364 * 10^{-15}$.

Note that an electric potential divided by an electric field would leave us with meters; that would tell us the amount of resistance experienced when a particle passes through a bound state.

The potential outside the well when there is a tau particle in the area is: $2.846842979 * 10^{-10}$; we divide by a coulomb and we have: $4.561145301 * 10^{-29}$. The tau generates its own electric field: $5.456280596 * 10^{-44}$: we divide, and we arrive at $1.196252309 * 10^{-15}$.

An electron probably lives forever; a muon lives for $2.1969811 * 10^{-6}$ s and a tau lives for $2.903 * 10^{-13}$ s .

There are two ways to solve for the momentum: $\hbar k$ or mv . We'll solve it both ways: then, with respect to a muon $k = \frac{(4.726490141*10^9)(\pi)}{2\pi} =$

$2.363245071 * 10^9$; we multiply by $\hbar$ and we have $p = 2.492211648 * 10^{-25}$. If we solve for mv : where $\frac{k}{v} = \frac{m}{\hbar}$ we have:

$\frac{(2.363245071*10^9)(2.1969811*10^{-6})(1.054571817*10^{-34})}{1.883531627*10^{-28}} =$

$2.906955111 * 10^{-3}$ meters, and so we have

$mv = 1.883531627 * 10^{-28} * \left(\frac{2.906955111*10^{-3}}{2.1969811*10^{-6}}\right) =$

$2.492211648 * 10^{-25}$.

So, the equations work out, and now we know how much our muon has moved (in and or with the well) in its lifetime—unless our muon is accelerating, and, consequently, k is changing.

Now, (where k is accelerating), we hold n constant: we're not adding any energy to our electric field; instead, a is decreasing: the breadth of our well is getting narrower, and the depth is increasing. Hence, the electric field is compressed. All the waves that once fit the well at $a = 2\pi$ are present in the compressed well, and, as such, the wave number has increased. Thus, if we know where the well is, then we can identify the location of the particle much better. (The well would exist wherever the vacuum is bombarded with values of k).

A muon in a well has no sense of direction, and, as such, it travels through time at the speed of light. A muon in a well where $a = 2\pi$ and $n = 4.726490141 * 10^9$ travels in and or with the well at a velocity of $1.323158907 * 10^3$. A muon that escapes the well, then, is traveling at the speed of light (as opposed to the speed of light squared) minus this amount: that velocity is 299779113 or $.999955486c$.

Therefore, if a muon should travel at this velocity throughout its lifetime (of $2.1969811 * 10^{-6}$ s), it should cover a distance of 658.609045436 meters.

If we allow k to be increasing and a to be decreasing, then all the information of the breadth is translated into information of the depth. We can use the above information to solve for the new

$a\ :\ \dfrac{n\pi\hbar}{mv} = a_2 = 2.773259457 * 10^{-5}$.

Then $k_2 = 5.354243602 * 10^{14}$ and $p = 5.646434404 * 10^{-20}$.

Then, a_2 is the distance that would be covered by the acceleration of the muon outside the well (or from the one level of a to the next).

The muon has accelerated from inside the well to outside the well (or an increasingly narrower well).

We can solve for the acceleration of the muon from one a to ever decreasing values of a .

We have $F = \dfrac{Q\,q}{4\,\epsilon_0\,r^2}$, where ϵ_0 (epsilon naught) is the capacitance of the vacuum.

Where $r = a$, we also have: $F = \dfrac{mv^2}{a}$ and so

$F = \dfrac{(1.883531627 * 10^{-28})(299779113^2)}{(2.773259457 * 10^{-5})^2} = 2.200871326 * 10^{-2}$.

Now, the electric potential of the external energy of the system (which is a function of an electro magnetic field) is $2.712218766 * 10^{-30}$.

(The electric potential of the internal energy of the system (the charge of the energy of the wave function) is $32^{.138122100} = 1.613966350$).

Then we have: $q = 1.60217663 * 10^{-19}$.

We solve for Q by taking the equation $V_q = \dfrac{Q}{4\pi\varepsilon_0 * a_2}$. $Q = 4.980293046 * 10^{-15}$.

Then: $a_3 = \sqrt{\dfrac{Q*q}{4\pi\varepsilon_0 * F}} = 1.805096231 * 10^{-11}$,

where a_3 is the minimum breadth or depth of the well.

Next, $\dfrac{c^2}{a_3} = 4.978987620 * 10^{27} \, \dfrac{m}{s^2}$.

Now, acceleration, $\vec{a}$, is equal to $\dfrac{F}{m}$, and, as such, a force divided by a mass equals acceleration even if the mass is not moving. Therefore, the force acting our muon, (from a_2), is

$2.200871326 * 10^{-2}$ and the force acting on our

muon, (from a_3) , is $1.406411362 * 10^4$.

We divide the force of a_3 by the force of a_2 and we get: $6.390248014 * 10^5$. That's the net force acting on a muon that is not moving for $2.1969811 * 10^{-6}$ s .

If the muon is accelerating at $\frac{F_{net}}{m}$ ($3.392694830 * 10^{33} \frac{m}{s^2}$), then, after $8.836410966 * 10^{-26}$ seconds, it would be traveling at $299792458 \frac{m}{s}$.

If the muon were traveling at the speed of light, (through both space and time), then, after traveling a distance of a_3 , it would be accelerating at $4.978987620 * 10^{27} \frac{m}{s^2}$.

We multiply the number of seconds it would take to reach the speed of light by the acceleration of a muon that would be traveling at the speed of light and we have: $439.963808048 \frac{m}{s}$.

So the muon would be traveling at that velocity through both space and time. Then the velocity through space or time alone would be 299792458 minus that: 299792018 .

We subtract the former velocity of the muon from the latter and we have: 12905 $\frac{m}{s}$.

Then 12905 $\frac{m}{s}$ * 2.1969811 * 10^{-6} s we have .028352041 m .

After so many seconds, then, the wave packet, or the muon that lives in the wave packet, would be ejected from the electric field. It would move at the speed of light if it weren't for its mass; nevertheless, because of the energy packed into the muon, and the muon's electric field, it continues to accelerate.

If it weren't still accelerating, after 2.1969811 * 10^{-6} s , the accelerating muon would travel an additional .028352041 m .

It doesn't, though, and so there must have been a delay between the conception of the muon and the point at which the muon decays into other particles.

We take the initial velocity of the muon (the observed value) and divide by the additional meters to arrive at the delay incorporated between the well and the muon's final destination.

That's a delay of $9.457643902 * 10^{-11}\ s$.

2

We'll solve, now, (as in the above), for the up quark: $E_n = \frac{\hbar^2 \pi^2 n^2}{2ma^2}$: the up quark would have a mass of $2.2 \frac{MeV}{c^2}$ or $3.921856227 * 10^{-30} \, kg$.

Then where an initial k would be 10430.37835, $n = 6.820216046 * 10^8$. Next, $\log_2 n = 29.345242200$.

Then we take $29^{.345242200} = 3.198022405$.

Then the wave function of the up quark must have an electric potential of $\frac{\hbar c k}{\frac{2}{3}C}$ where k would equal 3.198022405—and the charge of an up quark is two thirds the charge of an electron—so the electric potential is $1.079934708 * 10^{-44}$.

We have: $E_n + V = mc^2$; when it comes to an up quark, we have a potential outside the well of: $3.524788593 * 10^{-13}$ and an electric potential (of an electric field outside the well) of $3.764889272 * 10^{-32} \frac{J}{C}$.

We'll solve, now, for the radius of our up quark's electric field; the electric potential of our up quark's electric field is $1.079934708 * 10^{-44}$.

(So, the up quark would be very close to the source of the external electric field).

The charge of the electric field acting on our up quark would be $\frac{3.764889272 * 10^{-32} \frac{J}{C}}{\hbar c * 10430.37835 \ J}$. That would be $8.758802187 * 10^9 \ C$.

(Note that 10430.37835 is an arbitrary number representing the wave number, and through that, the energy, of an electric potential).

Now, the greater the charge of the external electric field, the higher the electric potential of the up quark.

Thus, the external electric field, (which tells us where the test charge is), and the electric field of the up quark would be a ways apart;

But they can't be, because the electric potential isn't very high.

Based on these numbers, the radius of the up quark's electric field would be: $r = \dfrac{|Q|}{V_q * 4\pi\varepsilon_0}$.

That would be $7.289156915 * 10^{63}$.

If that were the case, then the up quark would experience a very small electric potential with a humongous radius;

That radius, then, must be a function of an extradimensional region; we'll think, then, of r pointing in a direction perpendicular to the x and y plane. Thus r would lead to a z-coordinate that those of thus that live on the x and y plane couly only witness as a shadow to some extradimensional reality that contains the vector r . In order to get as close as possible to this z-dimension, we would have to travel a distance up and a distance over that would be greater than the distance of the vector r .

A combined length of 10^{63} , then, would tells us how far we'd have to go to get as close to r or the z-coordinate in question as possible.

If each dimension is a function of n bisections,

then we'd have $\log_2 7.289156915 * 10^{63} = r$.

That's 212.147221936 .

To get the actual distance of the electric potential from Q to q we follow the above steps for finding the radius of the electric potential Q and divide that r by the above r .

Where $r_Q = 74.432771913$ we have a distance

of 2.850185698 between the Q electric potenti

al and q respectively.

Now, let's solve for the location and momentum of an up quark that's undergoing an acceleration; (note that the unit of force, a newton, is equal to the force that would give a mass of one kilogram an acceleration of one meter per second per second; it's a measure of what would happen, and so the quark might be accelerating without going anywhere; other forces might be holding it in place).

Where $k = \frac{n\pi}{a}$, $n = 6.820216046 * 10^8$ and $a = 2\pi$, $k = 3.410108023 * 10^8$.

Then, where $\hbar k = mv$ we would hold the velocity of the up quark to be $9.169647243 * 10^3$.

Then, where $\bar{a}$, (acceleration), equals $\frac{v^2}{r}$, and $r = 2.850185698$, we have $\bar{a} = 2.950068503 * 10^7$.

Therefore, the velocity would be constantly changing, and, because the velocity is limited, the velocity must be cyclical, because it goes on changing forever. Hence, we'll be dealing with harmonic motion—(where the velocity of an object that is not moving would yield a change in wavenumber or a change in mass).

Then, where $\hat{H}$ is the Hamiltonian, $\hat{H} = \frac{\hat{p}^2}{2m} + \frac{1}{2}m\omega^2\hat{x}^2$; ($m\omega^2 = $ a spring constant, k , and the work or energy that took or would take place is

$\frac{1}{2}kx^2$, which is the potential energy, which is

force times distance, which is given by the integral

$V = \int_0^x kx' dx'$).

The Hamiltonian is the total energy of our system: we want our Hamiltonian to be the product of two factors, (otherwise, we can't isolate any variables) so we'll write it as the sum of two things

squared: $\frac{1}{2}m\omega^2\left(\hat{x}^2 + \frac{\hat{p}^2}{m^2\omega^2}\right)$ where $\frac{1}{2}m\omega^2 *$

$\frac{\hat{p}^2}{m^2\omega^2} = \frac{\hat{p}^2}{2m}$.

$\hat{x}$ and $\hat{p}$ are operators, and so they don't commute; (in the cross terms they change their order);

we factorize $a^2 + b^2 = (a - ib)(a + ib)$ and we

have $\hat{x}^2 + \hat{p}^2 + \frac{i}{m\omega}[\hat{x},\hat{p}]$ where $[\hat{x},\hat{p}] = i\hbar$.

($\frac{\hbar}{i}\frac{d\psi}{dx} = p$ and so $\frac{\hbar}{i}\frac{d}{dx} = \hat{p}$; $\hat{x}$ is x times

something—an operator is that which acts on

something; then we have: $(x)\left(\frac{h}{i}\frac{d}{dx}\right) - \left(\frac{h}{i}\frac{d}{dx}\right)(x)$: then, using the product rule, we have

$\frac{h}{i}\left[x\frac{d(\psi)}{dx} - x\frac{d(\psi)}{dx} - (\psi)\frac{dx}{dx}\right] = \frac{h}{i}(-\psi) = ih$).

The Hamiltonian becomes

$$\hat{H} = \frac{1}{2}m\omega^2\left(\hat{x}^2 + \frac{\hat{p}^2}{m^2\omega^2}\right)$$

where $\hat{x}^2 + \frac{\hat{p}^2}{m^2\omega^2} = \left(\hat{x} - \frac{i\hat{p}}{m\omega}\right)\left(\hat{x} + \frac{i\hat{p}}{m\omega}\right) + \frac{\hbar}{m\omega}$,

where $-\frac{\hbar}{m\omega}$ is the product of $\frac{i}{m\omega} * i\hbar$.

Then we have: $\hat{H} = \frac{1}{2}m\omega^2 V^\dagger V + \frac{1}{2}\hbar\omega$. Remember we are using operators, and so the cross terms don't work, which means we must add on the commutator.

Now, [V , V†] , where V† is V-dagger, is

$\left[\hat{x} + \frac{i\hat{p}}{m\omega}, \hat{x} - \frac{i\hat{p}}{m\omega}\right]$. We solve for the commuta tor (that which is in brackets) of that and we have

$$-\frac{i}{m\omega}[\hat{x},\hat{p}] + \frac{i}{m\omega}[\hat{p},\hat{x}] \rightarrow \frac{2\hbar}{m\omega} = [V, V^\dagger] \text{ and}$$

then we have: $[\sqrt{\frac{m\omega}{2\hbar}} V, \sqrt{\frac{m\omega}{2\hbar}} V^\dagger] = 1$.

We have $\hat{a} \equiv \sqrt{\frac{m\omega}{2\hbar}} V$ and $\hat{a}^\dagger = \sqrt{\frac{m\omega}{2\hbar}} V^\dagger$.

The equations that we've been searching for are:

$$\hat{a} = \sqrt{\frac{m\omega}{2\hbar}}\left(\hat{x} + \frac{i\hat{p}}{m\omega}\right) \quad \hat{a}^\dagger = \sqrt{\frac{m\omega}{2\hbar}}\left(\hat{x} - \frac{i\hat{p}}{m\omega}\right)$$

$$\hat{x} = \sqrt{\frac{\hbar}{2m\omega}}(\hat{a} + \hat{a}^\dagger) \quad \hat{p} = i\sqrt{\frac{m\omega\hbar}{2}}(\hat{a}^\dagger - a).$$

Now, we have $\sqrt{\frac{m\omega}{2\hbar}}$, and that has the unit m^{-1}.

That unit corresponds to Ψ^2, so we would have $A^2 e^{2ikx - 2i\omega t}$. (The i is part of the amplitude).

Then $m^{-1} = A^2 e^{2ikx - 2i\omega t}$.

If we take the derivative with respect to t, we have: $2i\omega e^{2ikx-2i\omega t}$.

Once we've been struck by a momentum, one unit of time passes from the time struck to the time observed: thus, we now have $\left(\frac{x_2-x_1}{t_2-t_1}\right)$ as opposed to $\frac{x}{t}$. Then we can write: $-4\omega^2 e^{2ikx-2i\omega t}$.

So, where the amplitude is equal to 1, we have a have the following: $\int |\Psi|^2 \, dx = 1$ and so we have $x = \frac{1}{|\Psi|^2}$. Then $|\Psi|^2 = \frac{1}{m}$. We also have the above, which is the second derivative of the wave function (and is, in and of itself, a wave function): $m = 1 / \frac{m}{s^2}$.

Then, m^{-1}, in the case of an accelerating wave function, (or, when it comes to harmonic motion), we have $m^{-1} = \frac{m}{s^2}$.

Now, the up quark was accelerating already,

and, when we struck it with a momentum, we accelerate that acceleration. We have $m^{-1} = \frac{m}{s^2} = \sqrt{\frac{\bar{m}\omega}{2\hbar}}$ where $\bar{m}$, in this case, refers to mass.

We multiply $\hat{x}$ by m^{-1} and we have a unit free $\hat{x}_2$.

We can also write out the units such that: $\frac{m}{s^2} * \hat{x} = \frac{m^2}{s^2} = v^2 = \hat{x}_2$;

We can also write $\frac{v^2}{r} = \frac{\hat{x}_2}{r}$; (note the velocity is the square root of $\hat{x}_2$ and r is the original value of $\hat{x}$).

Thus, a unit free value for $\hat{x}_2$ would correspond to an acceleration.

(Squaring $\sqrt{\frac{\bar{m}\omega}{2\hbar}}$ once is $\frac{1}{[V,V^\dagger]}$; we multiply that by $\hat{x}_2$ and $\sqrt{\dfrac{\hbar}{2m\omega}}$ and we get $\hat{x}_1$.

Therefore, it would be sufficient to square $\sqrt{\dfrac{\bar{m}\omega}{2\hbar}}$ once because the change in $\hat{x}$ accounts for any change in the commutator operators and any change in the commutator operators accounts for any change in $\hat{x}$).

So we have: $\hat{x}_1 = 2 * \dfrac{\bar{m}\omega}{2\hbar} * \sqrt{\dfrac{\hbar}{2\bar{m}\omega}} * \hat{x}_2$.

After two seconds, we'd multiply by $\sqrt{\dfrac{m\omega}{2\hbar}}$ And arrive at the new value of x.

3

What if we're dealing with a well with an energy of $8.987551787 * 10^{16}$ (299792458^2) ? If we stick with the up quark, and we solve for n (where $a = 2\pi$), we get $n = 1.592338465 * 10^{28}$.

Then $\log_2 n = 93.685133682$; that's the wave number of the potential in the well.

We take $93^{.685133682} = 22.318960855$; that's the left over wave number, and so the electric potential of the particle is $\frac{\hbar c k}{C}$: that's $6.606199725 * 10^{-6} \frac{J}{C}$ where the coulombs are two thirds the charge of an electron.

Then the potential outside the well is $mc^2 - E_n = V$; $V = (\sim) - 8.987551787 * 10^{16}$. Thus, the potential outside the well is a little less than 299792458^2, so the potential in the well cancels with the potential outside the well except for a tiny amount of potential that is negative outside the well and therefore would be potential inside the well. That's one plus an extremely tiny amount.

Then the electric potential outside the well would be negative and or potential inside the well:

$$\tilde{E} = \frac{(\sim) - 1}{1.068117753 * 10^{-19}} = (\sim) - 9.362263638 * 10^{18} \frac{J}{C}$$

(outside the well). That was $(\sim) - 1 = \frac{299792458^2}{mc^2 - E_n}$.

Where $\tilde{E} = \tilde{E}_0 \, e^{ikx}$, we divide $\frac{\tilde{E}}{\tilde{E}_0}$ and so,

$$e^{ikx} = \frac{-9.362263638 * 10^{18}}{6.606199725 * 10^{-6}} = -1.417193550 * 10^{24} \; ;$$

Now, we'll solve for the charge of our particle's wave packet: we divide the electric potential by the charge of the up quark in joules:

$$\frac{6.606199725 * 10^{-6} \frac{J}{C}}{1.068117753 * 10^{-19} J} = 6.184898347 * 10^{13} \frac{1}{C} \text{ and}$$

we have a charge of $1.616841448 * 10^{-14}$ C .

We'll solve for the radius of the absolute value of the electric field acting on that charge: $\tilde{E} =$

$\frac{|Q|}{4\pi\epsilon_0 r^2}$; $r = 3.939707567 * 10^{-12}$.

That is to say that a potential (in the well) is connected to an electric field that exists inside or outside the well.

We have $k = \frac{n\pi}{2\pi} = 7.961692325 * 10^{27}$ and

v would be $v = \frac{k\hbar}{m} = 2.140867960 * 10^{23}$:

(we divide by the speed of light squared—the maximum velocity of something moving with respect to both the speed of light through space and the speed of light through time—which would be possible if there were no reference point to measure our movement. Thus, because the velocity is in excess of the speed of light squared, there must be a reference point: hence, the division).

The new velocity is $2.382036856 * 10^6$ and

therefore the new k is $8.858577410 * 10^{10}$ and the new omega is:

$\omega = 299792458 * k$ so, we're interested in the

angular frequency of the space around the particle's wave packet—which has it's own angular frequency—but that angular frequency isn't a function of the spatial radiation that surrounds it. (We're interested in what would be harmonic motion between the wave packet and the spring like qualities of space interacting with the wave packet.

$\omega = 2.655734696 * 10^{19}$ (where $c = \frac{\omega}{k}$).

The particle is struck or affected by the $\tilde{E}$ −field (outside the well): the acceleration of the acceleration must go to zero; (the wave packet must conform to a new acceleration governed by an external force).

We can solve, then, for a new acceleration and a new velocity specific to where the particle is.

At an energy level of $8.987551787 * 10^{16}$,

we'll solve for $r_1 = 2 * \frac{m\omega}{2\hbar} * \sqrt{\frac{\hbar}{2m\omega}} * r_2 =$

$2.838237252 * 10^{-29} = r_2$, where r_1 is the magnitude or the distance from the origin, which was $3.939707567 * 10^{-12}$;

Then we have: $r_2 = \sqrt{\frac{m\omega}{2\hbar}} * 2 * \frac{m\omega}{2\hbar} * \sqrt{\frac{\hbar}{2m\omega}} * r_3$.

Now, changing the commutators changes the velocity of our object (on a spring, for example).

Then, where $\frac{v^2}{r_2} = a$, we have r_2 to be

$2.838237252 * 10^{-29}$. We have a to be a measure of a distance divided by an acceleration which tells us how fast that distance is accelerating.

We have: $\frac{r_1}{m\omega / 2\hbar}$ which is $7.977995499 * 10^{-36}$ s^2 .

and so $v_2 = 5.613355805 * 10^{-24}$.

The acceleration of r_3 is given by $\dfrac{r_2}{\sqrt{\dfrac{m\omega}{2\hbar}}^3}$

Next, $\dfrac{v^2}{r_3} = a$ where $r_3 = 5.747493599 * 10^{-53}$, $a = 8.178943531 * 10^{-65}$, and so, $v_3 = 6.856269072 * 10^{-59}$.

So, the acceleration(s) of the acceleration goes to zero.

4

We'll argue, now, that the momentum of a particle, such as the up quark, (contained by the gravity and the electric field of the earth), is inversely proportional to the momentum of the earth; we have:

$$m_1 \bar{v}_2 = \frac{1}{m_2 \bar{v}_1}$$, where m_1 is the mass of the up quark, and m_2 is the mass of the earth ($5.9722 * 10^{24}$).

Then v_1 which is a part of $\bar{v}_1$, is the speed of light divided by the initial velocity of the particle—which was $2.140867960 * 10^{23}$ or $2.382036856 * 10^6$ after dividing by multiples of 299792458 ; that's the speed of the particle through space and time—and so, to get the velocity through space alone, we divide the speed of light by this speed through space and time combined.

Note that, without a reference, an object, on the one hand, may not be moving at all, whereas, on the other hand, time stands still as the object moves through space at the speed of light and time at the speed of light. Thus, the lesser the velocity through space and time combined, the closer the velocity gets to the velocity through space or time alone. Thus, we have the velocity through space to

be $v_1 = 125.855507754$.

Then, to arrive at $\bar{v}_1$, we discount the speed of the orbit of the solar system around the center of the galaxy: we do that by dividing the speed of the earth around the sun by the speed of the solar system and the speed of the sun around the center of mass of the solar system divided by the speed of the solar system. That leaves us with the speed of the particle around the center of the galaxy, so we divide by the speed of the particle. Then we multiply by the speed of the rotation of the earth (which doesn't apply to the speed of the solar system around the galaxy)—and we have the net (initial) velocity of the particle.

That's $\dfrac{29784.8}{251000} * \dfrac{20400}{251000} * 465.1 * \dfrac{1}{125.855507754} =$

.035641135.

Then we have v_2 to be the velocity of the up quark after accelerating one unit (after affecting an electric potential): that was $5.613355805 * 10^{-24}$; then we divide one by that amount in order to arrive at the magnitude of the change of an incremental increase in meters that would amount to a velocity or derivative of $1 \frac{m}{s}$.

Then we take that velocity to be $5.613355805 * 10^{24}$. We divide by 299792458^2 to arrive at a velocity consistent with the maximum velocity of The speed of light. That is $6.245700654 * 10^7$.

Then we have: $\frac{29784.8}{251000} * \frac{20400}{251000} * 5.9772 * 10^{24} * 6.245700654 * 10^7 = 3.597434715 * 10^{30}$.

Then $m_1 \bar{v}_2 * m_2 \bar{v}_1 = .502847292$.

Now, we want to show that the momentum of the up quark apart from the earth (but equidistant from the sun) is inversely proportional to the momentum of a corresponding potential energy well.

That would be $m_1 = \frac{k\hbar}{v_1}$, so, where $k = 8.858577410 * 10^{10}$ and $v_1 = 125.855507754$, we hold the mass of the up quark apart from the mass of the earth to be $7.422802738 * 10^{-26}$;

Now, we have the kinetic energy: $KE = \frac{1}{2}mv^2$ or $\frac{\hbar^2 k^2}{2m}$; and we have the potential energy of the well and or the spring: $PE = \frac{1}{2}m\omega^2 x^2$.

Then, where the energy of the bound state is given by the negative kinetic energy in the well, that kinetic energy is potential energy outside the well. (The particle is still trapped inside the well, and the bound state exists outside the well—as a potential above the well).

We have $\frac{1}{m} = \frac{1}{2}\frac{1}{KE}v^2$; therefore, the new mass is a product of the kinetic energy inside the well, which would be all or part of the potential outside the well.

So, where KE is defined to be PE is defined to be 299792458^2 we have m_2—a part of the potential—to be $1.134819917 * 10^{13}$, where $v = v_1$.

Then, when it comes to $\bar{v}_1$, we're in outer space, so, whereas, before, when we divided the velocity of the revolutions of the solar system into the speed of the orbits of the earth and the sun in order to discount the velocity of the solar system around the galactic center, we must now divide by the speed of the orbits of the earth and the sun in order to discount them.

So, we have: $\bar{v}_1 = \frac{251000}{29784.8} * \frac{251000}{20400} *$

$125.855507754 = 1.296082362 * 10^4$.

Next we solve for $\bar{v}_2$, which is the velocity of the up quark after being affected by an electric field.
We have v_2 to be 4.799981215 . I got that by taking the velocity of the up quark after an ac

celeration of one unit ($5.613355805*10^{-24}$), in verting it, dividing by 299792458^2 , and then dividing 299792458 by that.

Then $\bar{v}_2 = 251000 * 4.799981215 = 1.204795285 * 10^6$.

Note that when the particle is in space, the affected or accelerated velocity doesn't exclude the velocity of the earth or the sun.

Moreover, we simply multiplied the velocity of the solar system by v_2 because the affected velocity is a product of itself and the velocity of the solar system.

Then $m_1 \bar{v}_2 * m_2 \bar{v}_1 = .013153482$.

Now, we'll define E_n to be 299792458 . The mass of the up quark would remain the same:

$m_1 = 3.921856227 * 10^{-30}$.

The mass of the earth would remain the same:

$m_2 = 5.9722 * 10^{24}$.

The initial velocity, then, would be where $n =$

$\sqrt{\frac{E_n 2ma^2}{\pi^2 \hbar^2}} = 9.196551739 * 10^{23}$; then $k = \frac{n\pi}{a} =$

$4.598275870 * 10^{23}$. Then v would be

$1.236458416 * 10^{19}$.

Now, gravitational potential— $\frac{J}{kg}$ —is equivalent to velocity squared: therefore, where $E = mc^2$ and c^2 is the gravitational potential of the mass in question, then, when motion is involved, we must multiply the mass in question by the number of joules per kilogram that would be or were taken out of the system.

Thus, where 299792458^2 is the energy (and the gravitational potential) when the system is at rest, (the mass isn't interacting with anything and or it's uniformly everywhere, so it must be equal to one), we divide an amount of energy (increments of 299792458) out of 299792458^2 (or any energy in excess of that) and we're left with a constant that would equal the number of joules in question.

Therefore, when the mass of the system shows up, we must multiply by the amount of energy (now a constant) equal to or greater than 299792458 that we took out of 299792458^2 . (If the energy of the system changed, the energy of the speed of light through space and time must've accounted for it).

So $v = \dfrac{\hbar k}{m*299792458} = 4.124381329 *$

10^{10} ; then dividing by 299798452 leaves us

with the velocity through space and time:

137.574552618.

(Note: the velocity through space would be

$\dfrac{299792458}{137.574552618} = 2.179127261 * 10^6$ and the

velocity through time would be $\dfrac{137.574552618}{2.179127261*10^6} =$

$6.313286750 * 10^{-5}$.)

Then $\bar{v}_1 = \dfrac{29784.8}{251000} * \dfrac{20400}{251000} * 465.1 *$

$\dfrac{1}{2.179127261*10^6} = 2.058453903 * 10^{-6}$.

Then v_2 would be a function of the accelera-

tion: that's where $r_2 = \dfrac{\frac{r_1}{m\omega}}{\hbar} * \dfrac{1}{\sqrt{\frac{\hbar}{2m\omega}}}$ and the acceleration is $\frac{\frac{r_1}{m\omega}}{\hbar}$; then, where $\frac{v^2}{r_2} = a$, we have $\tilde{v}_2 = 7.138042258 * 10^{-27}$.

That was where $r_1 = 3.939707567 * 10^{-12}$ which is the radius of the electric field acting on the up quark's wave packet; then we took the new k value from the velocity through space and time: 137.574552618 which led to the new omega: $1.533819730 * 10^{15}$.

Now, we want the magnitude of the change: so we take v_2 to be $\frac{7.138042258*10^{27}}{299792458^3} = 264.921378325$.

We divide 299792458 by that and we have: $1.131628032 * 10^6$.

$\bar{v}_2 = \frac{29784.8}{251000} * \frac{20400}{251000} * 1.131628032 * 10^6 = 1.091392864 * 10^4$.

Now we multiply the mass by the number of joules per kilogram we took out of the energy (and the gravitational potential). That's 299792458 . Then we divided by the velocity through space and time. (The velocity through space and time was a part of the gravitational potential that was absorbed by the above number (and therefore needed to be put back in).

Then $m_1 \bar{v}_2 * m_2 \bar{v}_1 * \dfrac{299792458}{137.574552618} =$

1.146649963 .

Now, we'll analyze the proportionality of the mass of an up quark in space and the mass associated with the particle's potential—where the kinetic energy in the well (negative kinetic energy) is equal to the particle's potential outside the well.

We want to show, then, that the particle's potential outside the well behaves like gravitational potential (energy per a corresponding mass).

We'll find that, whereas, on earth, we were multiplying the mass in question by the number of joules and or joules per kilogram that would have been removed from the system, in space, on the other hand, we're dividing the mass in question by the energy that would be taken out of the system; (that's because, in space, apart from the earth, the mass in question is a part of the particle's potential both inside and outside the well).

Then we have the mass of the up quark to be:

$m_1 = \frac{k\hbar * 299792458}{v_1}$ where, as on earth, k would

remain the same and v_1 would remain the same,

leaving us with $m_1 = 7.422802738 * 10^{-26}$.

(The division of v by an additional 299792458

discounted the multiplication of 299792458).

Then, for m_2 , we have $\frac{m}{299792458} = \frac{2KE}{v^2}$.

Then, where the velocity through space and

time (v^2) is 137.574552618 , we have $m_2 = $

1.306571836 $* 10^{15}$.

Then as in the former, we have: $\bar{v}_1 = $

1.296082362 $* 10^4$.

We multiply the affected or accelerated velocity of the particle times the velocity of the solar sys tem as in the former: $4.799981215 * 251000 = 1.204795285 * 10^6$.

Now, as mentioned in the above, the division of the former v ($1.236458416 * 10^{19}$) by 299792458^2 accounted for the multiplication of m by 299792458—so we don't need to divide m by 299792458 .

Then $m_1 \bar{v}_2 * m_2 \bar{v}_1 = 1.514422563$.

5

What if we're dealing (on earth) with an electron-neutrino and a well with an energy of 299792458^2 ?

We have: $E_n = \frac{\hbar^2 \pi^2 n^2}{2ma^2}$. We solve for n , (where $a = 2\pi$ and the mass of a neutrino is $1.25 * 10^{-37}$) and we get $n = 2.842788449 * 10^{24}$.

Then $\log_2 n = 81.233581020$; that's the change of the wave number and the wave number of the potential in the well.

We take $81^{.233581020} = 2.791167357$.

That's the left over wave number, and so the potential of the particle in the well is

E_0 =8.824350323 * 10^{-26} . That's $\hbar c k$.

An electron neutrino has no charge, but it is magnetized, and it travels as an electromagnetic wave; the spin (which yields a magnetic moment) generates a charge when the magnetized part of the particle and or the particle is in motion, but the new fields would take away from the field before them—and would be increasingly less than the fields before them, and so the charge in question would be limited if not negligible.

Therefore the electron neutrino potential will be considered gravitational instead of electric.

Then the potential outside the well (discounting the potential of the particle) is $mc^2 - E_n =$ V; $V = \sim - 8.987551787 * 10^{16}$.

Now, we know that $|\Psi|^2$, or $|e^{ikx}|^2$, is measured by the units of inverse meters.

Therefore, we can say that: $F = E_0 * |e^{ikx}|^2$,

where E_0 times the cosine or sine of the angle in question is the amplitude of the electron neutrino's wave packet.

So, $e^{ikx} = m^{-\frac{1}{2}}$ and so $e^{ikx} * e^{ikx} = m^{-1}$.

We can also say that $F_2 = V * \left|e^{ikx}\right|^2$. That's the force outside the well acting on the electron neutrino's wave packet.

In both cases, then, the amount of force would be equal to the amount of potential.

To arrive at the net force between the exterior of the well and the particle we divide:

That's:

$$\frac{F_2}{F_1} = \frac{V}{E_0} * \frac{\left|e^{ikx}\right|^2}{\left|e^{ikx}\right|^2} = F_3 = \sim -1.298423325 * 10^{16}$$.

That's a unit free number that relates two forces and or two gravitational potentials.

Now, we'll solve for the radius or the distance of the gravitational field existing between the electron neutrino and the earth.

We have $F_3 = G * \frac{m_1 m_2}{r^2}$ where G is the gravitational constant, m_1 is the mass of an electron neutrino ($1.25 * 10^{-37}$) and m_2 is the mass of the earth.

Then $r = 6.194651598 * 10^{-20}\ i$.

We take the magnitude, and we have $r_1 = 6.194651598 * 10^{-20}$.

We have $k = \frac{n\pi}{2\pi} = 1.421394225 * 10^{24}$ and where $v = \frac{k\hbar}{m}$ we have $v = 1.199169832 * 10^{27}$;

Then the velocity through space and time would be $v = 44.506002242$ and so $k = .052753641$ and $\omega = 2.347853654$.

Then the acceleration is $\frac{r_1 * \hbar}{m\omega} = 2.225932602 * 10^{-17}$.

Then $r_2 = \frac{r_1 * \hbar}{m\omega} * \frac{1}{\sqrt{2 * 1.25 * 10^{-37} * 2.347853654}} = 1.660655109 * 10^{-18}$.

So, where $\frac{v^2}{r} = a$ we have $v_2 =$

$6.079890088 * 10^{-18}$.

Then we take the magnitude of the change and we have: $\dfrac{6.079890088*10^{18}}{299792458^2} = 67.647900472$.

Then $v_2 = \dfrac{299792458}{67.646900472} = 4.431659459 * 10^6$.

Now, the neutrino passes through the earth: therefore the initial velocity is divided by the velocity of the solar system—the neutrino, then, is not related to the velocity of the earth or the sun.

So, $\bar{v}_1 = \dfrac{\frac{299792458}{44.5066002242}}{251000} = 26.836055794$.

Then v_2 , (the affected velocity), is multiplied by the velocity of the solar system, and we have:

$1.112346524 * 10^{12}$.

Then (on earth) we have:

Then $m_1 v_2 * m_2 v_1 = 22.303169704$.

We solve, now, for the mass of the particle apart from the earth: where $k = .052753641$ and

$v_1 = 6.736000604 * 10^6$ we have:

$m_1 = \frac{k\hbar}{v_1} = 8.258981302 * 10^{-43}$.

Now, neutrinos rarely interact with matter, and the walls of a well, hypothetically, would be made of matter, so the electron neutrino cannot be contained by the well for long. We're not dealing, then, with a particle that gets trapped by space or time—and so we consider the neutrino trapped indefinitely by space and time acting in unison.

We take the total wave number of the system (as opposed to the wave number of the particle passing through space alone) to be: $1.421394235 * 10^{24}$; then we calculated the value of omega from that times v_1 , which was $9.574512358 * 10^{30}$, which will tells us what the potential of the electron neutrino in the well would be if the electron neutrino could be trapped in the well.

We hold the mass, in space, of the electron neutrino to be $8.258981302 * 10^{-43}$.

We calculate the potential of the electron neutrino: $PE = \frac{1}{2} * m * \omega^2 * x^2$ where x in this case, is equal to the breadth of the well, which we've defined to be 2π .

So, $PE = 2.261898342 * 10^{26}$.

Then $m_2 = \frac{2KE*PE}{(v_1)^2} = 8.960665158 * 10^{29}$.

Now, because the electron neutrino is already passing through, the initial velocity, v_1, is already excluded by the affected or accelerated velocity. So, we have the initial velocity to be

$1\ \frac{m}{s}$ through space and time, or $299792458\ \frac{m}{s}$

Through space—which would exclude the velocity of light through time (since time, or the reference in question, at the speed of light, would cease to exist).

Then v_2, or the affected velocity through space, which is $4.710367196 * 10^5$, is multiplied by the velocity of the solar system, which leaves us with the net velocity through space and time.

Thus we have $\bar{v}_1 = 1\ \frac{m}{s}$ through space and

time and $\bar{v}_2 = 1.182302166 * 10^{11}\ \frac{m}{s}$ through

space and time.

Then $m_1 \bar{v}_1 * m_2 \bar{v}_2 = .087497414$.

6

What if we're dealing with a photon (on earth) and a well with an energy of 299792458^2 ?

We have: $E_n = \dfrac{\hbar^2 \pi^2 n^2}{2ma^2}$.

A photon has no mass, so we'll need to solve for the mass (a variable amount of energy) in terms of k .

We have: $m = \dfrac{k\hbar}{v} = \dfrac{KE2}{v^2}$.

So, the change in KE will correspond to a change in k that will yield m in terms of k .

Now, we solve for n by taking the mass of the system to be $\dfrac{KE2}{v^2} = 2$;

Hence, if the energy in a well remains constant, (and the breadth of the well remains constant) then as the wave number goes up, the amplitude must go down: so, when we hold the breadth of the well constant, too, the particle becomes that much more difficult to find.

Now, if we consider that with each addition of n , each a would be bisected, then it goes to follow that if we hold a constant, then we can say that the change in k , (where k is given by $\frac{n\pi}{a}$), is given by an accumulation of bisections, or $\log_2 k = n$. So, we're adding π to every half wave of the system. (We should note that k includes the other half of the wave, from the origin to $-a$).

Then, Δk is given by $\frac{n}{\Delta n}$, where Δn would be $\log_2 n = \Delta n$.

Therefore, $\Delta k = \frac{1.137115380 * 10^{43}}{\log_2 1.137115380 * 10^{43}} =$ $7.950283164 * 10^{40}$.

Then m_2 would be, $\frac{\Delta k \hbar}{v} = .027966496$ and if we hold the energy of the well constant, then KE must be $1.256751656 * 10^{15}$.

Thus, $\Delta KE = \frac{299792458^2}{1.256751656 * 10^{15}} = 71.514143209.$

That's the output of energy divided by the input of energy which is the derivative of the energy with respect to itself and or the total energy.

Then m in terms of k is $\frac{\Delta KE2}{v\hbar} = \bar{k} = 4.524025787 * 10^{27}$.

If the energy in the well is going down, (a is doubled for each change in n) then

$\log_2 n = k$ where k is the wave number of the potential in the well—meaning that k (and the energy of the well) only change at precise intervals that may or may not be integers of n . (As a gets bigger the change in k gets smaller because it stretches out).

Then $\log_2 n = 143.028286728$.

We take $143^{.028286728} = 1.150714019$.

That's the left over wave number, and so the potential of the particle (the energy of the wave packet outside the well) would be: $E = \hbar c k$ where $k = 1.150714019$ and so $E =$

$3.638013177 * 10^{-26}$.

We substitute $\frac{x \, m^{-1}}{k-constant}$ for kg (that were in the units making up the energy) and we got $x \, kg$ in terms of k . We have $1.645846543 * 10^{-52} \frac{m}{s^2}$.

(We wanted to divide the kilograms out of the energy. So, that was: $3.638013177 * 10^{-26} \left(\frac{1 \, m^{-1}}{4.524025787 * 10^{27}} \right) * m^2 * s^{-2}$).

Thus, the space around the photon is falling or accelerating toward the photon at that rate—which slows the photon in the same way that the earth slows down when it is falling or accelerating towards a falling object.

The velocity of the object is not changed by the earth falling toward it, and yet, the object shall get to the earth that much sooner. As such, the photon is traveling—or would appear to be traveling—through time.

Now, the circumference of a circle would be

equal to the wavelength of the wave: $2\pi r = \lambda$: (the radius of the wave is not necessarily the amplitude of the wave); we know that $\frac{v^2}{r} = \frac{m}{s^2}$, so, if we allow for a photon to be moving through space and time at $1\ \frac{m}{s}$ (which equates to the speed of light through space or time) then $1\ \frac{m^2}{s^2}$ over the radius of the universe is equal to the acceleration and or gravitational field strength of the photon; that would be $1\frac{m^2}{s^2} * \frac{1}{4.4*10^{26}\ m} = 2.272727272 * 10^{-27}\ \frac{m}{s^2}$.

We should note that the radius of the universe squared is approximately the same as the mass of ordinary matter in the universe $1 * 10^{53}\ kg$. That would make sense because the maximum velocity through space and time would not allow for a reference—so there would be nothing to define space or time. Mass, then, would be evenly distributed everywhere throughout the universe.

Now, if we take $g = \frac{GM}{r^2}$, where we substitute the mass in terms of the wave number (for this particular system), and we take r_2 to from $\frac{2\pi}{k} = \lambda$, so, where k was $\frac{n\pi}{2\pi}$, we have lambda to be $1.105109546 * 10^{-42}$ and so r_2 must be $1.758836469 * 10^{-43}$. Then g must be $4.769026414 * 10^{47}$.

That's the gravitational field strength or acceleration of the photon falling toward space, (as opposed to the gravitational field strength or acceleration of space falling toward the photon at a rate of $3.638013177 * 10^{-26} \frac{m}{s^2}$).

We multiply the two values of g and we have the gravitational field strength or acceleration of space in general and or overall—or, on the hand, we might say that we have gravitational field strength or acceleration of the wave function in general and or over all. We end up squaring the units, so we take the square root of the product, and we have: $1.31786395 * 10^{11}$.

We have $k = \frac{n\pi}{a} = 5.685576900 * 10^{42}$ and

Where $v = 299792458$, we have: $\omega = 1.704493074 * 10^{51}$.

The particle is struck or affected by the g-field (the gravitational field outside the well) and the acceleration of the acceleration (of the photon—which can mean changes in direction, amplitude, wavelength, and or frequency) will eventually stop; we can solve, then, for a new acceleration (and a new velocity) specific to where the particle is.

At an energy level of 299792458^2, we'll solve from $r_1 = \frac{m\omega}{\hbar}\sqrt{\frac{\hbar}{2m\omega}} * r_2$ for r_2. (We're working with a wave packet, so the former r_2 becomes the above r_1. That was $r_1 = 1.758836469 * 10^{-43}$. So, the new r_2 would be: Then, $r_2 = 9.198529027 * 10^{-73}$.

That's where $m_1 = 4.420841291 * 10^{-28}$

(So, we're taking the mass of the system and expressing it in terms of the wave number—that's m_1).

Now, we arrived at r_2 by taking $m = \frac{KE2}{v^2}$ which, in this case, left us with $m = 2$. So, we took m in terms of k to be

$$\frac{2*m^{-1}}{4.524025787*10^{27}} = 4.420841291 * 10^{-28} \ m^{-1} \ .$$

Then we take a to be: $\frac{m*(kg*m^2*s^{-1})}{kg*s^{-1}}$; now the kilograms included by Planck's constant are not defined: therefore we assume that they relate to our energy system and would be $4.524025787 * 10^{27} \ m^{-1}$.

(The kilograms of a defined mass in question, however, would be $\frac{m^{-1}}{4.524025787*10^{27}}$. Note that m^{-1} absorbs whatever kilograms were in the nu merator).
(Thus, the unit shifts when we translate a particular mass to inverse meters because we have to divide the existing mass out of the inverse meters).

(Thus we have $x \ kg = \frac{x \ m^{-1}}{C}$ where C is 4.524025787 * 10^{27}).

Then the acceleration, substituting

4.420841291 * 10^{-28} in for the kilograms, is:

$$\frac{1.758836469 * 10^{-43} * \hbar}{4.420841291 * 10^{-28} * 1.704493074 * 10^{51}} = 2.461509170 *$$

$10^{-101} \ kg * m^4$.

Now, the kilograms are no longer contained by the h-bar constant, and, furthermore, no preexisting kilograms are given, so we must take the mass with respect to the kilograms to be

$4524025787 * 10^{27} \ m^{-1}$, and we have: $a =$

$1.113593096 * 10^{-73} \ m^3$.

Then $a = 4.811075547 * 10^{-25}$.

We multiply by $\frac{1}{\sqrt{\frac{\hbar}{2m\omega}}}$ where $m =$

$4.420841291 * 10^{-28}$ and $\omega = 1.704493074 *$

10^{51} .

So, $r_2 = 5.751341775 * 10^4 \ m$;

Then where $v_2^2 = a * r_2$, and we have $v_2 = 1.663434925 * 10^{-10} \ m$. So, we can divide by any number of seconds in question and arrive at the accelerated part of the velocity of the wave packet—which would be that much less than the speed of light.

These units can be thought of as $\frac{m*s^2}{s^2}$ meaning the number of seconds (such as earth seconds) that will fit into the number of seconds (such as velocity seconds) decreases by a factor of meters times (velocity seconds squared) per seconds squared.

Then, where $m_1 = \frac{2*m^{-1}}{4.524025787*10^{27}} = 4.420841291 * 10^{-28}$, we multiply by the mass of the earth: we have: .002640215 .

Next, the initial velocity would be 299792458.

The affected velocity would be: $1.663439925 * 10^{-10}$.

Then we multiply the affected velocity by the velocity of the solar system and we have: $m_1 \bar{v}_2 * m_2 \bar{v}_1 = 33.047568167$.

7

What if we're dealing with a photon apart from the earth at an energy level of 299792458^2 ?

Where $m_2 = \frac{KE2}{v^2}$, we have $m_2 = 2$.

Now, with respect to a photon on earth, we used the mass in and or of the well to solve for m_1 (in terms of k) with respect to this m_2 .

So, to arrive at the new m_1 , we must solve for the new wave number of the photon.

We've included the 2 already as the mass m_2 . Therefore the mass in terms of k is simply $4.524025787 * 10^{27} \ m^{-1}$.

Now: $k = \frac{v_1 m_1}{\hbar}$: $k = \frac{299792458 \ ms^{-1}}{1.054571817 * 10^{-34} \ m^2 s^{-1}} =$

$2.842788449 * 10^{42} \ m^{-1}$.

We use that value of k to solve for m_1, which is 1 —which means that the mass of the photon is uniformly spread out everywhere, and, as such, the particle itself would have no mass.

Now, if the mass were spread out uniformly everywhere, then we'd be traveling at the speed of light through both space and time—or 299792458^2.

So, there are no reference points—but we're not always traveling through space and time together—the faster we travel through space the slower we travel through time and the faster we travel through time the slower we travel through space.

The mass associated with the well, then, would be our reference point; that keeps us from traveling faster than the speed of light. So, to arrive at v_1 we would divide by 299792458.

So, we have $\bar{m}_1 = 1 \ kg$.

We have $\bar{m}_2$ to be $2 \ kg$.

We hold r_1 to be $1.758836469 * 10^{-43}$.

Then the acceleration $\frac{r_1 \hbar}{m\omega}$ is $a =$ $2.176401045 * 10^{-74} \ m^2$. That's where m was $\frac{(1) \ m^{-1}}{4.524025787}$. So the acceleration is equal to

$1.475263043 * 10^{-37}$ m .

We multiply by $\dfrac{1}{\sqrt{\dfrac{\hbar}{2m\omega}}}$ and we have $r_2 =$

$4.408142016 * 10^{-8}$, which has no units—or meters per meters—which would be the number of meters gained after traveling so many meters.

Then $v_2 = \sqrt{a * r_2} = 8.064222842 * 10^{-23}$ $m^{\frac{1}{2}}$.

Then $v_2 = 6.503169005 * 10^{-45}$.

We take the magnitude of the change of an incremental increase in meters and we arrive at what would be a negative velocity of

$6.503169005 * 10^{45}$ $\dfrac{m}{s} * s$.

We divide that by the 299792458^5 and we have

$v_2 = 63.499666554$ $\dfrac{m}{s} * s$.

Then we take that velocity (that would be velocity through space and time) and divide it out of

299792458 $\dfrac{m}{s}$ and we have $4.721165862 *$

10^6 s . We multiply by the velocity of the solar

system and we have $1.185012631 * 10^{12}$ m.

Now, we haven't always multiplied all the units out, such as the velocity, because the various velocities were all part of the same velocity and or velocity vector. In the above, however, we're dealing with different units.

To continue, traveling at the velocity of light is velocity through space and time (the speed of light squared) such that, from this timeless perspective, everything is evenly distributed.

Note that velocities through space and time are multiples that can be interchanged—we just divide or multiply by multiples of the speed of light. This is possible because the speed of light cannot travel faster than the speed of light—or, if it could, it would be a function of velocity through space or time.

Now we divide v_2 out of 299792458 and we have $\bar{v}_2$ to be $3.952776661 * 10^3$ s.

So we would have 1 $kg * 299792458 * 2$ $kg *$

$\frac{x}{3.95277661*10^3 \ s} = 1$. Hence, if the momentum of a

photon is inversely proportional to the momentum of the energy of and or associated with a potential well (as opposed to a planet, for example) then the meters in question must be: $4.129364308 * 10^{-13}$.

That way we would wind up with $m_1 \bar{v}_2 * m_2 \bar{v}_1 = 1$.

8

We can create a differential equation such that

$\frac{d^2 h}{du^2} - 2u\frac{dh}{du} + (\varepsilon - 1)h = 0$. That's where $\varepsilon = \frac{2E}{\hbar\omega}$. We got this equation, first of all, from the Schrodinger equation:

$-\frac{\hbar^2}{2m}\frac{d^2\varphi(x)}{dx^2} + \frac{1}{2}m\omega^2 x^2 \varphi(x) = E\varphi(x)$. That's Where $x = au$ where a^2 is a constant given by A combination of the constants of the problem: $a^2 = \frac{\hbar}{m\omega}$ and where u , then, is a unit free coordinate.

Now, where $\frac{1}{a^2} = k$, we take $\frac{\hbar^2}{m} * \frac{m\omega}{\hbar} = \hbar\omega$

and we also have $m\omega^2 * \frac{\hbar}{m\omega}$ so we have:

$$-\frac{1}{2}\hbar\omega * \frac{d^2\varphi}{du^2} + \frac{1}{2}\hbar\omega u^2 \varphi = E\varphi \ .$$

Then we multiply both sides by $\frac{2}{m\omega}$ and we have: $-\frac{d^2\varphi}{du^2} + u^2\varphi = \frac{2E}{\hbar\omega}\varphi \ .$

So, $\frac{d^2\varphi}{du^2} = (u^2 - \varepsilon)\varphi \ .$

Now, we'll define $\varphi(u) = h(u)\, e^{\frac{-u^2}{2}}$.

We hope that $h(u)$ (a function) is a proxy for φ and this function $h(u)$ won't diverge.

(The differential equation for φ implies a differential equation for h).

(We can assume that $h(u)$ might contain $e^{\frac{u^2}{2}}$ such that $e^{-\frac{u^2}{2}}$ cancels out. Then we plug in $h(u)$. We have: $\frac{d^2 h}{du^2} = (u^2 - \varepsilon)h \ .$

We take the derivative of the u term to arrive at our differential equation—(containing all the derivatives of h). (We use the chain rule on the differential u term and we add the h term to complete the polynomial we're creating. We arrive at $\frac{d^2h}{du^2} - 2u\frac{dh}{du} + (\varepsilon - 1)h = 0$).

This equation is a polynomial, which means it approaches a point in x, and, as such, with respect to x, doesn't diverge.

So, we had the series: $h(u) = \sum_{k=0}^{\infty} a_k u^k$.

We imagined this series with a term

$a_j u^j + a_{j+1} u^{j+1} + a_{j+2} u^{j+2} \ldots$. Then, if we take

two derivatives to end up with u^j, then we must

have started with $a_{j+2} u^{j+2}$.

Then: $\frac{d^2h}{du^2}$ would be $(j+2)(j+1) * a_{j+2} u^j$.

Next we look at the $-2u\frac{dh}{du}$ term. If we

start out with h we have: $a_{j-1} u^{j-1}$.

Then we differentiate with respect to u which leaves us with $\frac{dh}{du}$.

We remember the series, where we have:

$h(u) = \sum_{k=0}^{\infty} a_k u^k$, so, if we took the derivative of

h , (with respect to u), we'd have $\frac{dh}{du} = k *$

$a_{k-1} u^{k-1}$. Note that j is a substitute for k .

Now we multiply by u and so, for the series to hold, u must be $(j+1)$ — and so we have

$(j+1)(j-1) a_j u^j$ So, we brought down the $(j+1)$ and the $(j-1)$ and that left us with j by itself. We carry the negative two from the differential equation and we have: $-2j a_j u^j$.

(A number $(j+1)$ plus a number $(j-1)$ and divided by two leaves us with the original number j . So, j contains two terms—one told us about the derivative of h and one told us about what that derivative must've been.

Another way to look at this is to say that if you have $-2u \frac{dh}{du}$ and you integrate once to arrive at what you must have started with you have:

$-\frac{2}{2}u^2 * h$. So you divided by two and you got the correct term—which includes the average of

$(j+1)+(j-1)$.

For the $(\varepsilon-1)h$ term, no derivatives were taken in order to arrive at h ; so we started out with $a_j u^j$. That leaves us with: $(\varepsilon-1)a_j u^j$.

So all of the differential equation is:

$\sum_{j=0}^{\infty} [\,(j+2)(j+1)a_{j+2} - 2ja_j + (\varepsilon-1)a_j\,] * u^j = 0$. So the terms in the brackets must be equal to 0 . Therefore, we have:

$(j+2)(j+1)a_{j+2} = (2j+1-\varepsilon)a_j$.

So, $a_{j+2} = \frac{2j+1-\varepsilon}{(j+2)(j+1)} * a_j$.

In order for this series to be both a polynomial and normalizeable, it must truncate. The series must terminate as it goes to zero.

Then, where $\frac{a_{j+2}}{a_j} = \frac{2j+1-\varepsilon}{(j+2)(j+1)}$, ε must be equal to $2j + 1$.

So, when this series terminates, $a_{j+2} = 0$.

Thus, the last coefficient that can exist is a_j and then, (because $a_{j+2} = 0$), you go down by two's. So you'd have: $h(u) = a_j u^j + a_{j-2} u^{j-2} + a_{j-4} u^{j-4} \ldots$.

Then, where $j = n$, we have: $E = \frac{\hbar\omega}{2}(\varepsilon)$ so we have: $E = \frac{\hbar\omega}{2}(2n + 1)$.

Then, we have: $E_n = \hbar\omega\left(n + \frac{1}{2}\right)$.

That's the total energy $E_{\{n\}}$ of the bound state—whereas $E_n = \frac{\hbar^2 n^2 \pi^2}{2m*a^2}$ is the negative kinetic energy of the bound state or the potential in the well or the potential of the bound state.

Thus, the total energy of the bound state $E_{\{n\}}$ minus the potential of the bound state E_n equals the total potential energy of the system. Thus, the total energy of the bound state is greater than the total potential energy of the system—hence, a particle in the potential would be trapped by that potential and would yield a bound state.

We can also observe that $\hbar\omega\left(n + \frac{1}{2}\right)$ is greater than mc^2 so mc^2 must only govern energy outside the bound state.

Now, as n increases, the total energy of the bound state gets bigger faster than the kinetic potential of the bound state and the total potential energy of the system, V_{total}, so the total potential of the system gets bigger at an accelerating rate as it approaches the total energy of the likewise accelerated bound state (where n is getting bigger and bigger).

Thus, if the entire universe were a bound state, the energy of that bound state would get bigger at an accelerated rate (with respect to the kinetic potential of the bound state).

Thus the total potential energy of the system accelerates until it equals the total energy of the bound state minus the kinetic potential of the bound state.

We might say, then, that the total potential energy of the system plus the kinetic potential of the system forces an extradimensional region into existence—which is the remainder of the total energy of the bound state once n is greater than one.

So, what happens is that, if the universe were a bound state, then, as that bound state expands by a factor of n , a second state and or a second reality is acknowledged.

Now, the ordinary matter of the universe is approximately $1.5 * 10^{53}$ kg . Thus, the ordinary energy of the universe, (before it cancels with the ordinary matter of the universe times the gravitational potential of that universe) is: $1.348132768 * 10^{70}$.

So, the total energy of the bound state is

$\hbar\omega \left(n + \frac{1}{2}\right) = 3.370331920 * 10^{70}$, where $n = 2$.

So, the total potential energy of the system is

$1.348132768 * 10^{70} - \frac{\hbar^2 n^2 \pi^2}{2ma^2}$.

So we would have: $\frac{\hbar^2 n^2 \pi^2}{2*1.5*10^{53}*(4.399239670*10^{26})^2} =$

$7.561980581 * 10^{-174}$, where a is the radius of the universe.

But, the area of the wave function that would be harboring a mass of $1.5 * 10^{53}$ (in otherwise empty space, such that the mass would be spread out uniformly) is equal to one—so the mass of the area in question would be: $\frac{1}{1.5*10^{53}}$.

The area of the wave function that would be harboring a singular particle, such as an electron—in otherwise empty space, such that the mass would be spread out uniformly with respect to an amplitude—would be equal to the mass of that electron divided by 1 (or the total area of the wave function).

Furthermore, a , the radius of the universe, (when it comes to a particle trapped in a well) is directly related to the energy of that particle, and inversely related to the energy of a condensation of particles. Therefore a bunch of particles at a point increases the energy at that point (and decreases the energy everywhere else)—and thereby reduces (or would reduce) the breadth of the well.

Now, when we invert the mass and the breadth of the well, we have the kinetic potential of the bound state to be: $6.372774739 * 10^{39}$.

Now, how much more did the total energy of the bound state go up with respect to the kinetic potential of the bound state? We divide the former by the latter, and we have: $5.288641225 * 10^{30}$.

Hence, the total potential energy of the system went up by that much—(where $n = 2$).

Then, relatively speaking, where the universe is a potential well, the total potential of the bound state goes to infinity, and the kinetic potential—or the kinetic energy—goes to zero (albeit at its own pace). The amount of energy in the universe, then, is increasing and the kinetic energy is decreasing. The rate at which the kinetic potential energy is decreasing: $5.288641225 * 10^{30}$: would be the rate (per seconds if we add an additional axis to measure time) that a universe with a bound state and a secondary state would expand. (If kinetic potential energy, over all, is decreasing, (as the total potential of the bound state goes to infinity), then the area of the potential goes up).

Now, the kinetic potential, or the kinetic energy bounded by the well, is increasing or expanding as n goes to infinity; thus, the kinetic potential is both going to zero (with respect to the total energy of the bound state and the total potential) and going to infinity (with respect to everything else). So, we can quantize two different infinities with respect to each other—where both infinities are equally valid. We have a region or reality, then, that would function as a second bound state and or a second universe (where the universe is considered a bound state). In fact, as n goes to infinity, we have, for each value of n greater than 1, a different bound state and a different region. (If n equaled one, which it must if the universe, as a bound state, exists as a bound state, the total potential of the bound state would exist as a ground state, and, therefore, without a change, a second bound state and or the total potential of the system would go undetected).

Now, a blackbody is an idealized object that absorbs all electromagnetic radiation and emits thermal radiation in a continuous spectrum based on its temperature.

Stars are like blackbodies because they absorb all photons that reach them and emit heat in proportion to their temperature.

(Stars absorb radiation because their hot gases are opaque).

A potential well does not emit a spectrum of radiation: it's not a perfect absorber. It can only absorb scattered states (finite quantized amounts of energy) that absorb enough energy at the bound state—such that they become a bound state.

Therefore a potential well cannot include a blackbody.

Because the blackbody must exist outside the potential well, the capacitance, with respect to Planck's constant (which deals with frequency) and or Planck's reduced constant h-bar (which deals with angular frequency) is a function of the potential energy well.

So the blackbody—a region outside our universal (potential) well, absorbs antimatter emitted or transferred to the blackbody from the well—(the particle trapped in the well implies the existence of its antimatter partner)—yields a finite energy level per a finite region defined by the pressure experienced by the blackbody.

Note that this region would exist outside both the well and the blackbody: it would be the universe we experience—or the total energy of a bound state of or as the system in question.

Thus, we have three regions: the universal well; the blackbody; and the universe we experience—which is the total energy of the bound state and or the region created by any number of bound states.

Now, the universe we experience is a function of a blackbody defined by discrete energy levels and the universal well—which is a function of our universe defined as the total energy of a bound state. Thus, the universe we experience is a finite region determined by the thermal radiation and or spectrum of the blackbody and the mass and energy contained in this our universal well.

The total energy of this finite region would be the sum of the kinetic and potential energies of the universal region plus the energy determined by the antimatter (the blackbody region) plus the amount energy determined by the amount of pressure exerted on the blackbody—which is a function of the total potential of the system.

That's: $\frac{1}{2}mv^2 + \frac{1}{2}m\omega_1^2 x^2 + \frac{1}{2}mc^2 + m\omega_2^2 x_2^2$,

where $\frac{1}{2}mc^2$ is the energy of the antimatter of the blackbody, ω_1 is $\frac{2\pi}{4.34786832*10^{17}}$ (two pi divided by the age of the universe) and ω_2 is the negative reciprocal of $\frac{mv}{\hbar} * c$; we took the nega

tive reciprocal because, in effect, we added a dimension when we considered the pressure experienced by the blackbody to be a function of the pressure exerted on the blackbody by our universal potential well (when antimatter is transferred from the universal potential well to the blackbody). So $m\omega_2^2$ = force per unit length—so that length would be equal to one; then the energy of the surface pressure experienced by the spring would be equal to $m\omega_2^2 x_2^2$ where $x = 1$.

To be clear, then: we have a universe defined as the total energy of a bound state; we have a blackbody that absorbs the antimatter half of the particle trapped inside the well; and we have the universal potential well.

So, where the mass of the universe is $1.5 * 10^{53}$, the radius (x) of the universe is $4.399239670 * 10^{26}$, the age of the universe is $4.34786832 * 10^{17}$, (and where we take ω_2^2 to be $\left(\frac{1}{-kc}\right)^2$, we have the total energy of our bound state or the total energy of our finite region to be

$3.779805306 * 10^{72}$.

Now, where a bound state is a region where a trapped particle is likely to be found, we divide the total energy of our finite region by the former total energy of the bound state (note that the bound state is a function of the potential, and, as such, it exists outside the potential) and we have the amount of energy that is leaking into our universal potential well (the total energy of the bound state) from this finite region.

Thus, we would exist in a bound state that can be defined by that which it absorbs or that which leaks into the universal potential well that generates the bound state: that's a unit free element that tells us the size or amount of a finite energy state.

That's 112.149348958.

We might conclude, then, that *a* universe we experience, (the region where the blackbody spectrum is released), is the region created by the gap between the total energy of the bound state and the total potential energy of the system.

9

We know that antimatter exists through the equation: $E^2 = m^2c^4 + p^2c^2$, (where m is the rest mass), which tells the magnitude or length of the energy vector. The root tells us that there must be positive and negative solutions—where the negative solutions would leave us with antimatter.

Thus, when investigating the magnetic moment, or the current times the area of a region—defined, in this case, by the area of the wave packet—then we apply a dimensionless constant to account for differences in what we calculate and what we observe. This constant, or g-factor, depends on the mass, charge, and angular momentum of the particle.

Thus, when investigating the current of a particle we apply a g-factor to account for the doubling of the mass—a mass that, in and of itself, has become a part of a vacuum or blackbody (and the radiation of that vacuum or blackbody).

When measuring energy, then, $E = \frac{p^2}{2\bar{m}}$, we

must account for the antimatter of the vacuum—and so we multiply the mass by two. When dealing with mass as a function of kinetic and potential energies—we must divide the total mass by two. That's why we have the lowest state of the well

and or the oscillator to be $\frac{\hbar\omega}{2}$. (The anti-particle

and the particle have been separated)

If we define a photon with right hand helicity, (the north pole of the spin goes in the same direction as the direction of motion), to be negative or anti-photons, and a photon with left hand helicity, (the north pole of the spin points in the opposite direction of the direction of motion), to be positive photons, then we'll assign the one photon to a blackbody and the other photon to a potential energy well—such that the blackbody absorbs and emits right handed or anti-photons, that, in turn, are absorbed by potential energy wells which would emit both right handed or negative photons and left handed or positive photons.

We can imagine that photons, instead of propagating through space as we know it, propagate up and down one after the other—producing a wave that constructs a group velocity out of photons that only move in one direction—up and down.

Now, when a well emits or shifts a photon the anti-photon goes into a blackbody; the momentum of the anti-photon generates enough energy to radiate another anti-photon that, in turn, goes straight into a well, which, at that, emits another anti-photon.

A left handed photon is emitted and reabsorbed by the well, generating extra-dimensional amplitudes that form a left handed wave.

So, we have the total energy of the bound state (the potential plus the kinetic potential) minus the bound state equals potential:

That's $\hbar\omega \left(n + \frac{1}{2}\right) - \frac{\hbar^2 n^2 \pi^2}{2ma^2} = V$.

We hold the phase velocity to be $v = \frac{\omega}{k}$ and the group velocity to be $v_g = \frac{d\omega}{dk}$.

Now, if we live in a bound state that is a function of the energy of the universal well that comprises our universe—then the blackbody and or the well emit left handed photons out into the universe and or directly into the bound state and or the universe we experience. So the propagation of an electromagnetic wave would be a function of the phase velocity or one dimensional left handed photons (points that represent extradimensional regions) and two dimensional right handed photons and or wave packets that fill up the space in the universe we experience as they shift back and forth between that space and the potential well or blackbody that comprises the kinetic potential of our universe.

So: right handed photon: group velocity photons that exist between the potential and the blackbody and are comprised of two or more dimensions;

And the left handed photon: phase velocity photons that, as one dimensional points, represent the intersection of our region and extradimensional regions (determined by the various heights of the wave).

Note that the left handed photon would not be observed unless it coincided with the right handed photon's wave packet.

Now, there are two differing measurements of the peak or dominant level of microwave background radiation. The peak radiation level when measured by frequency intervals is $1.6 * 10^{11}\ Hz$ which corresponds to a wavelength of $1.873702863 * 10^{-3}\ m$.

The peak radiation level when measured by wavelength intervals (which relate Wiens constant relating the wavelength and temperature of blackbody radiation to the temperature of the universe) is $1.063215270 * 10^{-3}$ m which corresponds to a frequency of $2.819677881 * 10^{11}$ Hz .

We'll show why that is. If we draw a triangle, then, within the first quadrant of a circle, where the vertical axis is a measure of angular frequency (omega) and the horizontal axis is a measure of inverse meters (k), we can solve for the radius and the angle from the origin.

Now, when a particle passes through a bound state and or over a potential well, there is a phase change that amounts to and or accounts for a delay in the journey of the particle.

Thus, we have a triangle that corresponds to the lesser frequency interval, which means that:

$$\omega = 2\pi f = 1.005309649 * 10^{12} \ s^{-1} \text{ and}$$

$$k = \frac{2\pi}{\lambda} = 3.353352034 * 10^3 \ .$$

Then $r = \sqrt{\omega^2 + k^2}$ and so $r \sim 1.005309649 * 10^{12}$ m^{-1} .

Now, we use the inverse sine function to arrive at the angle or phase change, which is $\theta \sim \frac{\pi}{2}$.

Now the arclength would be $m = r * \theta$ where m is the arclength, r is the radius, and θ is the angle in radians.

Then $m = 1.579136704 * 10^{12}\ m^{-1}$.

Then we solve for the new omega that is a function of the addition of m to k which leaves us with a new value for k that is $1.579136707 * 10^{12}$.

That's $\omega = \sqrt{r^2 - k^2} = 9.733868826 * 10^7 i$.

We take the magnitude to be the square root of the imaginary part squared and we have:

$\omega = 9.733868826 * 10^7$.

Then we divide this new value of omega—(determined by the measure of frequency intervals in the cosmic microwave background radiation—the CMB)—and divide it by the inverse meters associated with the measurement of the CMB with respect to wavelength intervals determined by the temperature of the universe: $\frac{9.733868826 * 10^7}{5.909607851 * 10^3} = 1.647126014 * 10^4\ \frac{m}{s}$.

We divide the measure of the frequency interval peak by the measure of the wavenumber with respect to the wavelength interval peak and we have a velocity of $1.701144432 * 10^8$.
Then we take the velocity with respect to the new value of omega (counting the delay) and the peak wavenumber with respect to the wavelength

interval and subtract: $(1.701144432 * 10^8 -$
$1.647126014 * 10^4 = 1.700979719 * 10^8$)

Next, we divide the frequency with respect to the wavelength interval by the velocity above and we have: $1.041550231 * 10^4 \quad m^{-1}$.

Now, we subtract the wavenumber with respect to the frequency interval and we have: $7.062150535 * 10^3$.

So the delay experienced by the frequency style of measurement amounted to the above—which is very close to the wavenumber we had in the first place: $5.909607851 * 10^3$.

Note that the longer the delay (the longer the photon is affected by the black body) the greater the temperature of the photon—(the blackbody doesn't stop absorbing more and more energy)—and this heat or excess energy is transferred to a potential well—making it possible for the particle(s) in the well to escape.

So, when passing through a bound state (and or a blackbody) there is a delay that amounts to the greater value of the wavenumber with respect to measurements devoted to the wavelength of the CMB. The blackbody expresses this value of k , and the well, in turn, expresses a lesser value k .

Now, if blackbodies and potential energy wells are functions of each other, we can solve for the phase and group velocities between them.

Thus, we have: $\omega_1 = 1.005309649 * 10^{12}$ and $\omega_2 = 1.7716558863 * 10^{12}$.

We also have $k_1 = 3.353352034 * 10^3$ and $k_2 = 5.90960751 * 10^3$.

Now, the phase velocity must be the greater of the two in order to keep up with the group velocity. Then the phase velocity must be: $\frac{\omega_2}{k_1} = 5.283238518 * 10^8$.

Then the group velocity, or the second derivative, would be $\frac{\omega_2 - \omega_1}{k_2 - k_1} = 299792458$.

What is the relationship between the wavelengths of the group and phase velocities? If the group velocity is measured by the up and down coincidental motion of photons implanted in space, what, then, is the radius of those photons and or their wave packets? Does that have any bearing on the wavelengths of the phase velocity, and, if so, what's moving faster—the photons of the group velocity or the waves of the phase velocity?

We can note, hereof, that the faster the waves of the phase velocity are moving, the easier it is to find the photon on the group velocity wave; the slower the phase velocity is moving, the more difficult it would be to discover the photon coincidental with the phase velocity.

What is the radius between the photons that make up the group velocity? Certainly, it should be smaller than the wavelength of the group velocity—after all, we're only allowing for up and down motion! Now, if we take current density, in one dimension, to be frequency, and if we're only allowing for up and down motion, then each photon must have a radius extending to another photon; we know, however, that a photon is smaller than the wave that contains it; and we know, in space, that group velocity nodes coincide with phase velocity nodes.

Therefore, the radius of a photon traveling at group velocity must be less than or equal to one-fourth the wavelength of the group velocity wave.

(The radius of the photons that make up the circumference of a circle that defines the wavelength of a wave at intervals of four must be less than or equal to one-fourth of the circumference. A circumference and or interval of $2\pi r$, then, must allow for a radius less than or equal to $\frac{2\pi r}{4}$). The radius of a photon that makes up a group velocity wave is therefore less than or equal to $\frac{\pi r}{2}$.

Now, the wavelength of the phase velocity should be less than the wavelength of the group velocity: indeed, there should be at least one complete wave per half of the group velocity wave—otherwise the group and phase waves would be the same. Furthermore, the wavelength of the phase velocity should be less than or equal to $2r$, where r is the radius of a photon that comprises the group velocity. (That's at least one phase velocity wave—and two photon radii—per half group velocity wave).

(The phase velocity wave over 2 must be greater than or equal to the radius of an individual photon or else there would be no room for the required phase velocity wave. So we have $\frac{1}{4}$ of our group velocity wave, which corresponds to at least one half a phase velocity wave, and we divide by two, which ensures that we have one phase velocity wave between two group velocity photons).

So, as we know, the change in the phase velocity, Δv_p , is the group velocity—more on that in a moment.

We want to know, for now, if the phase velocity changes within the boundaries of the group velocity. So we take the radius and or amplitude of a portion of the group velocity and call it $\omega_2 - \omega_1$. (All things being equal—if the wavelength remains the same—the change in frequency is the same thing as the change in amplitude). That's the vertical portion of a right triangle. The horizontal portion would be the radius (from the origin) of a particular photon and or photon wave packet, (corresponding to a quantity of energy expelled by a black body) which would be the greater value of k. The hypotenuse, then, yields the change in the phase velocity. (If we hold omega constant, then the change in the wave number is the change in the velocity).

Then we can solve for the specific radius and or amplitude of a specific phase velocity wave:

$$\omega_n = \sqrt{\Delta v_p^2 - (k_2 - k_1)^2}.$$

That's where the specific phase velocity wave is given by $k_2 - k_1$.

Thus, ω_n is a variable quantity that depends on the specific parameters of the system. That is to say that the phase velocity is subject to change within the boundaries of the group velocity.

Now, $\Delta v_p = v_g$, but v_g is defined at c. So, v_g is red-shifted according to the amplitude given by $v_g - \omega_n$ (where ω_n would be the change in amplitude of v_p).

It goes to follow that v_p slows down (it doesn't travel as far) corresponding to the red-shift of v_g .

Now, we want to show that the phase velocity photons exert pressure on the group velocity wave: then we'll solve for a force $\frac{\hbar k}{t}$ that repels space.

Now, time, at the speed of light, functions like a delta function: $\int_{-\infty}^{\infty} \delta(t)\, dt = 1$.

Given the properties of the delta function, such that everywhere that $t = 0$, the delta function is infinite, and everywhere that $t \neq 0$ the delta function is equal to zero, we must analyze the above integral with respect to time.

In the above integral, in order for the area or the time span of the delta function to be equal to one—a number without units—the delta function must be equal, in the above case, to an angular frequency. That way the units cancel.

Therefore, $\delta(t)$ is a function of inverse seconds.

Thus, given the properties of the delta function, if the time is equal to zero, then the angular frequency is infinite—and if time is equal to some number other than zero, then the angular frequency is equal to 0 .

As such, we can say that the delta function, as a function describing both time and the speed of light—(t , for the outside observer, would equal zero at the speed of light, and the angular frequency would go to infinity; for the inside observer, t would be infinite, and the angular frequency would go to zero)—leaves us any number of inverse seconds with respect to that delta function

The force exerted by photons, then, would be equal to the momentum of the photons per second— $\frac{\hbar k}{t}$, or $\hbar k * \delta(t)$.

Now, if we imagine a circle whose diameter is the radius of an even bigger circle, then the apparent brightness and or intensity of the electromagnetic wave within the smaller circle would be that apparent brightness times $\frac{1}{d^2}$ (or one over the ra

dius of the bigger circle squared), which would be

equal to $\frac{\pi^2}{4}$ if the circumference of the smaller

circle were equal to two. (That's $\frac{1}{d^2} = \frac{\pi^2}{4}$ is $\frac{2}{\pi} =$

d where $\pi * d = 2$).

Now, where the intensity varies by $\frac{1}{d^2}$, we can say that, at a distance of d^2 , the apparent brightness or intensity is equal to $1\ kg * m^2 * s^{-2}$ per second. Now, where kg and t are equal to 1 , (one joule, at rest, is accelerated by a factor of one second) we hold the intensity to be $\frac{1}{d^2} * kg * m^2 * s^{-3}$.

Where the intensity is measured in $kg * s^{-2}$ per second, (and $\Delta t = 1$) we can multiply by

the wave number and get the pressure and or energy density on the circumference of the smaller circle per second.

If we stretch the circle out into a wave with a wavelength of 2 , then we can multiply the intensity by the wavelength and arrive at a force.

The force delivered by the photons is equal to the momentum delivered by the photons:

$\frac{\hbar k}{t}$ where $t = 1\ s$.

Thus a force (amounting to the momentum) expresses itself along the walls of the circumference.

Now, pressure is defined as the force divided by the area—so, in this case, so far, we have a force expressing pressure on the boundaries of the circumference of the circle.

That force is delivered by photons—and so the phase velocity and the photons associated with it exert pressure on the space around them, (like a spring), and form group velocity wavelengths.

When we stretch the circle out into a wave with a wavelength of 2, we can multiply by the wave number to arrive at the pressure exerted by the photons on the line and or path of the group velocity photon(s).

The pressure is the force divided by the area: so,

we have: $\frac{\hbar k}{t} * \frac{1}{\pi r^2}$ where r is the radius of the smaller circle—which is $\frac{1}{\pi}$.

So the pressure, P, is $\frac{kg}{m\, s^2}$.

If we multiply the pressure by the wavelength, (which, for us, is still equal to 2) then we have the units of a spring constant. So what is the spring constant with respect to potential wells, blackbody radiation, (and the difference in measuring the peak microwave background radiation)?

So the wave number corresponding to this difference would be $k_2 - k_1 = 2.556255817 * 10^3$.

Then the force acting on the walls of our circle would be: $2.695755342 * 10^{-31}$. We multiply by one over the area and then again by the wavelength—we have the spring constant. That's $1.693793035 * 10^{-30}$.

Next, we take the spring constant, ($m * \omega^2$) and solve for the mass in order to solve for the energy in order to solve for the amplitude. Now, when starting at rest, the amplitude would be one half of the wavelength of a stretched out line (or spring) that is formed by the walls of our circle.

(The wavelength would be the initial displacement and or the maximum amplitude divided by the average amplitude—where the average amplitude is equal the maximum amplitude divided by the number of oscillations of our spring—where the stretch of one oscillation would be greater than the wavelength).

So, if everything but the spring constant and the amplitude remains constant (as it relates to our circle with a circumference of 2) we can show that the difference in the peak value of the microwave background radiation is a function of this spring constant.

We take our value for k and solve for the angular frequency: $7.663462147 * 10^{11}$.

Then the mass is $2.884099967 * 10^{-54}$.

Then $E = 2.592099781 * 10^{-37}$.

Then we get the amplitude (where the potential energy is at a maximum) from the potential energy equation—$E = \frac{1}{2} m\omega^2 x^2$:

The amplitude (x) is $5.532363204 * 10^{-4}$.

We multiply that by Plank's constant and the wave number and we have angular momentum: $1.956437565 * 10^{-34}$.

The energy associated with the lesser frequency ($\hbar\omega$) is $1.060171223 * 10^{-22}$.

We divide the angular momentum by that and we get the number of seconds that would make our joules equal to that angular momentum: $1.845397726 * 10^{-12}$ s .

The angular momentum divided by the greater angular frequency energy gives us $1.047153784 * 10^{-12} \ s^{-1}$.

So, the seconds needed to make the angular momentum equal to the energy is the difference in seconds: $7.982439420 * 10^{-13}$.

Now, to arrive at the number of seconds that would make the joules associated with the lesser angular frequency equal to the joules associated

with the greater angular frequency we have: $s *$

$1.060171223 * 10^{-22} = 2\pi * 2.819677881$

$10^{11} * \hbar$.

So $s = 1.762298676$.

(We basically just divided the greater angular frequency by the smaller angular frequency).

Now we divide the time it takes to make the angular momentum equal to the energy by the time it takes the lesser angular frequency energy to match up with the greater angular frequency energy and we have the delay experienced by the phase velocity with respect to the group velocity—which means that the wavelength of the phase velocity must've gotten bigger—(since we know that, in a vacuum, the phase velocity would be equal to the group velocity: that is to say that a point traveling on a phase velocity wave (although it may be moving faster or slower than the speed of light) would reach the end of the group velocity wavelength at the same time as the group velocity wave).

That delay would be: $4.529561038 * 10^{-13}$.

That would be seconds per seconds.

Next, the greater the pressure that the phase velocity applies to the group velocity the greater the group velocity wave will gain energy and contract. Thus, the pressure and or the angular momentum experienced by the group velocity add energy to both the group velocity and the phase velocity—which would increase the energy of the wave while, at the same time, increasing the wavelength (due to the elasticity of the group velocity wave).

www.ingramcontent.com/pod-product-compliance
Lightning Source LLC
Chambersburg PA
CBHW031445210526
45464CB00005B/2335